Memorize the Periodic Table

The Fast and Easy Way to Memorize Chemical Elements

2nd Edition

Kyle Buchanan & Dean Roller

ISBN: 0987564625
ISBN-13: 978-0987564627

To Rach, Monte and Lexy

CONTENTS

INTRODUCTION

You are about to memorize the chemical elements of the periodic table faster than you ever thought possible.

How do I know this? Because I've experienced exactly the same thing.

When I was younger I had to learn a lot of new information too. Whether it was chemical element names for Chemistry, quotes from Shakespeare for English, or the theories of marketing at college, it all had to be known by heart.

And so, I did what everyone else did. I went over the material again and again. And again. And eventually it would stick, for a short time at least. It won't be a revelation to you that learning in that way is boring and hard. It's no fun at all. Repetition is dull, dull, dull.

The turning point came when I saw someone memorize a deck of playing cards. Unless you count cards at a casino it's not an especially practical skill. But I couldn't help thinking how easily I could study and learn new information, if only I had a memory like that.

It turns out so called memory 'experts' aren't any different to you or me. Their brains are exactly the same. They're not any smarter and their memories aren't any better than yours or mine. What!? It's true. The only difference is they've learnt how to harness the potential of their memory using simple 'mnemonic' techniques. (Hint: Ditch the 'm' when trying to say that word! It is pronounced 'nemonic').

I spent years of school and college beating my head repeatedly against various textbooks, when I could have been memorizing that same information quickly and effectively? And it wouldn't have been boring? And I'd actually be able to remember it all now? Doh!

Bottom line, memorizing any information effectively is based on mental imagery. These mnemonic techniques have been around for over two thousand years, and there's 50 years of university research which proves their validity.

It's an exhilarating feeling when you unearth a talent you didn't know you had. Imagine how you'd feel if one day you suddenly realized you could paint like Monet or Picasso. That's what it was like when I discovered how easily I could remember anything. I memorized the periodic table, presidents, prime ministers and anything else, all because it was so easy. It was fun! I know, it sounds weird, right?

The primary mistake when trying to memorize is to use repetition and rote learning. People think it's the only way to remember new information. But it's not the only way, and it's rarely the most effective way either.

I became obsessed with researching mnemonic techniques and teaching other people how to use their memory properly.

In "Memorize the Periodic Table" we've gone a step further than simply teaching you a technique. Rather than explain how to create visual memory stories which your memory can grab onto, we've written the stories for you.

By simply visualizing each story and image as you read it, your amazing memory will do the rest. Each image will remind you of a chemical element, and will link to the element following on the periodic table.

You will be astonished at how quickly you're able to memorize the entire periodic table.

Enjoy discovering how amazing your memory is.

2 • HOW TO USE THIS BOOK

The golden rule for using this book is:

Create a clear mental picture of each image described.

It really is that simple. You don't need to try and 'remember' anything. Imagine each object, action or picture described as if you were seeing it right in front of you, and then trust your memory. Don't just read the words; actually visualize the image they're describing. As you get used to doing this you'll be stunned at how easily you'll be able to recall everything you visualized.

Each word you want to memorize is represented by an image. Sometimes this is easy to do. For example, if I see the word 'Gold', I automatically picture a shiny gold colored bar of the precious metal.

But some elements are less easily visualized. The

name 'Manganese' doesn't represent a physical object you immediately see in your mind. So what if you picture your knees with mangos on them? 'Mango knees' remind you instantly of the word 'Manganese'.

By substituting words for images in this way, you'll be able to picture every element name you want to memorize.

To link each name/image with the one following, we'll associate them using a ridiculous story. In the world of memory mnemonics, this is called the Link and Story Method. The stories are silly and exaggerated to help you remember them.

When you clearly imagine the visual story that links each separate picture, you'll link each word together and keep them in the correct order.

Again, *always clearly visualize everything described*, and your memory will naturally do the rest for you with surprisingly little effort.

The Link and Story Method used in this book will enable you to recite all the elements of the periodic table in order. If you wish to also quickly recall the atomic number of particular elements, you can read the final chapter on Remembering Atomic Numbers.

You'll quickly demonstrate to yourself how much quicker this method of memorizing is than repetition and rote learning. But you will still need to complete some mental rehearsal of your memory story to ensure the information is permanently transferred to your long term memory.

How often should you revise your newly acquired knowledge?

It's recommended you replay and revise your 'mental movie' five times. Ideally, first do this ten minutes after learning the story; then one day afterwards; one week later; one month later; and again in three to six months' time. Simply press 'play' on your mental movie and ensure you can still clearly see each mental image.

And that's how easy you'll find memorizing the periodic table. Create a clear mental picture of every image described, and run through your visual story five times in the months ahead.

Trust me, if you're anything like me, you'll run through your story many more times in the days ahead because you'll be so excited at how easily you can do it.

3 • HYDROGEN to NEON (1-10)

Let's begin by picturing a typical poster or chart of the periodic table. There are many small, colorful squares, each with a name, number and symbol of an element, and together they create a large irregular shape.

This image will act as an anchor in your memory, holding down the chain of images which link together all the elements.

Now we'll take that colorful poster and attach it to the first element. Picture that poster of the periodic table and imagine it's wrapped around a water hydrant.

Why a water hydrant?

1. Hydrogen

The 1st element in the periodic table is Hydrogen.

Hydrogen sounds similar to *hydrant* and that's how you're going to remember it. Picture a water hydrant you see on the sidewalk. It's short, stubby, red, and looks strong. The hydrant is like a little man with a small hat on top and stubby arms sticking out the side.

Imagine that hydrant with the chart of the periodic table wrapped around it. When you think of the chart of the periodic table, you'll picture it wrapped around a water hydrant. Because hydrant sounds similar to hydrogen, you'll know the 1st element in the table is Hydrogen.

2. Helium

The 2nd element is Helium. If you're like me, when you think of *Helium*, you automatically think of a *helium balloon*. When you let it go, it's the type that floats up into the sky. Now imagine an enormous helium balloon. Make it the size of a car and picture it attached to the water hydrant. Because the helium balloon is so big and has so much lifting power, it starts to lift the water hydrant up off the sidewalk. Together they slowly float up into the air and away into the sky. Now, when you visualize the helium balloon floating upwards, you'll know the 2nd element is Helium.

3. Lithium

The 3rd element is Lithium. *Lithium* sounds a bit like "*lithp*". People that have a lisp – a type of speech impediment – aren't able to pronounce "lisp" and say "lithp". Let's pretend the large helium balloon has a lisp. It also has a small hole in it, causing the balloon to slowly deflate. Usually a balloon with a hole in it will make a slow "ssss" sound, but because this balloon has a lisp or "lithp", it makes a "thhh" sound. Visualize the large balloon slowly deflating making a "thhh" sound. When you think of the balloon's "lithp", you'll be reminded of the 3rd element, Lithium.

4. Beryllium

The 4th element is Beryllium. If you say *Beryllium* slowly, it sounds like *"bee really yum"*. Picture your slowly deflating balloon. Imagine an enormous bumble bee lands on the balloon. The bee is the size of a football and has bright yellow and black stripes and buzzes loudly. The bee licks the balloon to have a taste and says, "that's really yum!" It really likes the taste of the balloon. When you picture the bee licking the balloon, you'll think, "bee really yum", and be reminded of the 4th element, Beryllium.

5. Boron

The 5th element is Boron. We can break up the word *Boron* into *"bore"* and *"on"*. The word "bore" can mean to drill a hole. Picture now the bee, after tasting the balloon. It uses its stinger, pierces the balloon and starts to spin around in a drilling motion. The bee has landed on the balloon, tasted it, and now it's started to "bore on" the balloon. When you picture the bee begin to bore on the balloon, you'll remember the 5th element, Boron.

6. Carbon

The 6th element is Carbon, which sounds like "*car bomb*". This is going to be an interesting one. After the balloon has been drilled or "bored on" by the bee, imagine the balloon suddenly explodes. Because the balloon is holding up the hydrant, the hydrant plummets out of the sky and lands on an army jeep. Visualize in your head what this would look like. As the hydrant is quite heavy and it has fallen from a great height, it makes an enormous "bang" as it smashes into the jeep. Bits of broken jeep go flying everywhere and it shocks everyone walking by. People cry, "Oh my gosh, it's a car bomb!" When you imagine the shattered jeep in your mind, you'll think "car bomb" and relate that back to the 6th element, Carbon.

7. Nitrogen

The 7th element on the periodic table is Nitrogen. If we say *Nitrogen* slowly and break it up, we get "*night row gen*". In this instance, we'll say "gen" is short for general, like an army general. Now, the general is the owner of the army jeep just crushed by the plummeting hydrant. Fortunately, the jeep was empty at the time because the general who owns it was out night rowing. He enjoys a bit of exercise on the river, usually in the hours of darkness. Now visualize him rowing his rowboat at night. Since he's well known for night rowing, the rest of the army calls him the "night row gen". When you imagine the "night row gen", you'll obviously be reminded of the 7th element, Nitrogen.

8. Oxygen

Oxygen is the 8th element on the periodic table. For *Oxygen*, picture an *oxygen mask*, same as the type that falls out of the ceiling for an airplane emergency. Now imagine the night row general, Nitrogen, finishing his rowing session. It was a particularly intense and exhausting session and he can barely stand up. He's completely exhausted and crawls out of his rowboat and back on to land. He needs to use an oxygen mask to get extra oxygen and air into his lungs to help him recover. When you picture the "night row gen" gasping and sucking from that oxygen mask, you'll be reminded of the 8th element, Oxygen.

9. Fluorine

The 9th element is Fluorine. *Fluorine* sounds similar to *fluoride*, like fluoride toothpaste or the fluoride gel dentists put on your teeth. The gel helps strengthen your teeth and it has an interesting taste and a weird texture as well, because it's like a paste that gets smeared onto your teeth. Imagine now the oxygen mask the general has been sucking oxygen from is now actually filled with fluoride gel. Picture him sucking in oxygen and then suddenly getting a mouthful of fluoride gel. All of his teeth are coated with it and he starts to spit it out. When you imagine him getting that mouthful of fluoride, you'll be reminded of the 9th element, Fluorine.

10. Neon

The 10th element is Neon. *Neon* is the chemical that goes into *neon signs*. Neon signs are the incredibly bright and colorful signs you see lit up at night. Imagine what Las Vegas looks like with all of those neon casino signs, blinking on and off, flashing with really bright and colorful images. Now imagine once the general spits the fluoride out of his mouth, it's made his teeth glow like neon. Picture him speaking, and you can see his teeth are all glowing brightly with different colors – blue and yellow and green and pink. When you visualize that, you'll be reminded of Neon, the 10th element.

Summary

Alright, that's the first 10 elements of the periodic table. I'm sure you can easily remember them, but quickly revise your memory story now by replaying it and "watching" it in your mind.

1. Picture a poster of the periodic table. The poster of the periodic table is wrapped around a hydrant, which reminds you of *Hydrogen*.

2. The hydrant starts to float up into the sky because it's got a huge helium balloon attached to it, reminding you of *Helium*.

3. When it gets high up in the sky, the balloon starts to slowly deflate and makes a "thhh" sound, because it has a "lithp", giving you *Lithium*.

4. Then you see a huge bumblebee land on the balloon, lick it and declare it's "really yum". Bee, really yum – *Beryllium*.

5. The bee starts to bore on the balloon with its stinger, making you recall *Boron*.

6. The balloon explodes, the hydrant plummets to earth and lands on an army jeep with an enormous bang. Everyone thinks there's been a car bomb. "Car bomb" sounds like *Carbon*.

7. Luckily, the jeep was empty as the general who drives it was out night rowing. He's the "night row gen" – *Nitrogen*.

8. The night row gen had a very intense rowing workout and uses an oxygen mask to help him recover. The oxygen mask makes you think of *Oxygen*.

9. Inside the oxygen mask was fluoride gel, reminding you of *Fluorine*.

10. The fluoride gel made the general's teeth glow like a neon sign – *Neon*.

You've now memorized the first ten elements of the periodic table. How easy was that?

4 • SODIUM to CALCIUM (11-20)

You're travelling well, so let's move on to the next ten elements. We'll continue on with our memory story from where we left off, with Neon.

11. Sodium

The 11th element on the periodic table is Sodium. *Sodium* sounds similar to soda or *soda bottle*. The general, the night row gen, is naturally disturbed by his teeth glowing various neon colors. He sees a piece of a broken glass soda bottle on the ground. He picks it up and tries to scrape his teeth and remove the glowing neon color. It's quite a sharp piece of soda bottle and he's extremely careful. Clearly visualize him cleaning his teeth with a piece of soda bottle, and you'll be reminded of Sodium.

12. Magnesium

The 12th element is Magnesium. While he's scraping

at his teeth with the bit of soda bottle, the night row gen notices a mag (magazine) on the ground. Strangely, it's full of pictures of knees. It is a *mag of knees*. Picture in your mind all of those anatomical images of different knees. Mag of knees reminds us of Magnesium, which is the 12th element.

13. Aluminum

Aluminum is the 13th element. Imagine the night row gen is about to bend over and pick up the mag of knees off the ground. Before he can reach it, he's suddenly hit from above with an enormous avalanche of *aluminum cans*. There are thousands of Coke, Pepsi and empty soda cans. They all land on his head and take him by surprise. Picture the unlucky general buried under a mountain of aluminum cans and you'll remember the 13th element, Aluminum.

14. Silicon

The next element is Silicon, the 14th element on the periodic table. Now, imagine the general stands up and shakes off the last remaining aluminum cans. He looks around to see where they came from. He sees a very *silly con* (convict) who had mistaken him for a policeman and had thrown the avalanche of aluminum cans at him. Picture this silly con dressed in prison stripes but he looks like a clown at the same time. He's got on a large red nose, a silly wig and oversized shoes. He's definitely a silly con, and reminds you of the 14th element, Silicon.

15. Phosphorus

The 15th element is Phosphorus. This is a challenging name, but let's go with "*fossil for us*". Imagine the silly con is on the run from the police with a stolen fossil. Visualize the silly con as he runs home to his family and bursts through the door, crying out "fossil for us." He's carrying an immense fossil. Imagine it's a dinosaur fossil and the silly con is struggling under its weight. Again, clearly picture him bursting through the door and screaming "fossil for us" to his family and you'll recall the 15th element, Phosphorus.

16. Sulfur

The 16th element is Sulfur, which sounds like "*sold fur*". Imagine the silly con has burst into his family home with the fossil, and his family actually sells fur, animal furs. Picture the inside of the family house with signs advertising furs for sale and signs that say "Sold Fur" on furs no longer for sale. The family is promoting how good they are at selling fur. Yes, it's a stretch, but visualize the "sold fur" signs really strongly in your mind and you will remember the 16th element, Sulfur.

17. Chlorine

Chlorine is the 17th element. *Chlorine* sounds a bit like "claw in". Imagine you've looked around and seen all those "sold fur" signs. You also notice how all of the furs are hanging on the walls. Imagine furs of bears, lions, tigers or animals that had big claws. The furs of those animals are stuck on the wall hanging there because they have their claw in the wall. Picture each animal fur with its "claw in" the wall, and you'll never forget the 17th element, Chlorine.

18. Argon

The 18th element number eighteen is Argon. If you break *Argon* up into two parts it sounds like "*R gone*". After you have looked at the wall of the furs hanging on the walls with claws in it, imagine one particular fur is quite an odd shape. It's been cut into the shape of the letter R. Clearly imagine the R shaped fur hanging on the wall and then watch it suddenly fall off the wall and disappear. The R is gone. R gone, reminding you of the 18th element, Argon.

19. Potassium

Potassium is the 19th element. If you break up the first part of the name *Potassium*, you get "*pot ass*". Imagine the R shaped fur has fallen into a large pot. The pot is the shape of a donkey, which is also known as an ass. You know what a donkey looks like, so imagine now the pot is the shape of what we'll call an ass. Pot ass will remind you of the 19th element, Potassium.

20. Calcium

The 20th element is Calcium. *Calcium* sounds similar to "*Colosseum*". The Colosseum is the half ruined remains of an immense stone stadium in Rome where gladiators and animals once fought each other. Picture the inside of the pot ass. There is a mini Colosseum in there. There are crowds screaming for blood while tiny gladiators are fighting each other. Colosseum equals Calcium, the 20th element.

Summary

It's time for you to rewind, replay and review your mental memory movie.

11. The night row general picked up a piece of broken soda bottle to scrape the neon off his teeth. The soda bottle reminds you of *Sodium*.

12. While he's leaning over to pick up that broken soda bottle he notices a magazine full of pictures of knees. "Mag of knees" helps you remember *Magnesium*.

13. As the night row gen is about to pick up the mag, he gets buried with an avalanche of aluminum cans. Aluminum cans represent *Aluminum*.

14. He stands up to see where those aluminum cans came from and sees a very silly looking con. Silly con – *Silicon*.

15. The silly con turns and runs under the weight of a large fossil. He bursts into his family home crying out "Fossil for us" reminding you of *Phosphorus*.

16. In your mental image of his family's home you can see "Sold fur" signs hanging on the walls. "Sold fur" gives you *Sulfur*.

17. There are also a lot of animal furs hanging on the walls. They are hanging on the wall because they have their claws in it. "Claw in" sounds like *Chlorine*.

18. One of the furs is cut in the shape of the letter R and it suddenly falls off the wall and disappears. "R gone" gives you *Argon*.

19. Where has it gone? It has fallen into a funny pot shaped like an ass or donkey. "Pot ass" reminds you of the element *Potassium*.

20. Finally, imagine yourself looking into the pot ass and inside it's like a miniature Colosseum with crowds screaming and gladiators fighting. "Colosseum" sounds similar to *Calcium*.

5   SCANDIUM to ZINC (21-30)

Alright, we are up to the 21st element. The last scene in our story contained an image of a miniature Colosseum to remind you of calcium.

21. Scandium

Scandium is the 21st element. *Scandium* reminds me of *Scandinavia* which is how it got the name scandium. When I think of Scandinavia, I think of *Vikings*. Go back to your mental image of the Colosseum with all the gladiators fighting. Instead of Roman gladiators, make them Vikings from Scandinavia. The Vikings combatants look fierce. When you picture these Vikings fighting, imagine their blond flowing hair and their horned helmets, and shields and swords in their arms. They're all slashing away at each other. The Vikings remind you of Scandinavia which in turn prompts you to recall the 21st element, Scandium.

22. Titanium

The 22nd element is Titanium. *Titanium* sounds similar to *Titanic*. Visualize all of the Vikings who've just been fighting each other. After winning their battles, the Vikings go back to their ship but instead of a classic Viking ship it is actually the infamous ocean liner, the Titanic. Imagine the very old large ocean liner and it's filled with Vikings. When you picture the Titanic swarming with Vikings you'll remember the 22nd element, Titanium.

23. Vanadium

The 23rd element is Vanadium. *Vanadium* sounds like *"van aid"*. Because the Titanic is so large and there are a lot of wounded Vikings on board, they have an aid van. It's like an ambulance, and is driving around on the deck of the Titanic treating all of the wounded. It's an aid van or a van aid. Visualize an ambulance and change it around. Remember it is a "van aid" and you will remember the 23rd element, Vanadium.

24. Chromium

The 24th element is Chromium. *Chromium* sounds like *chrome*. Imagine the van aid is stopping to help some wounded Viking gladiators. The doctor jumps out of the vain aid and he is made completely of chrome. Chrome is a very shiny, silvery and reflective metal. The doctor is made completely of chrome. Imagine him reflecting everything around him. He is very shiny and when you look at him, you can actually see yourself in his reflection. The doctor made of chrome will remind you of the 24th element, Chromium.

25. Manganese

The 25th element is Manganese. *Manganese* sounds like *mango knees*. Imagine the chrome doctor who's giving first aid treatment to each Viking. The initial treatment is a mango to put on their knees – mango knees. Imagine the doctor giving each Viking a big ripe mango and they slice it open and put half a mango on each knee. "Mango knees" will remind you of the 25th element, Manganese.

26. Iron

Iron is the 26th element. Picture again the Vikings with mangoes on their knees. The next stage of their strange medical treatment is to use a heavy iron, like a *clothes iron* you use to iron shirts. The doctor uses it to squash the mango onto their knee and to heat it up. See the doctor with his large clothes iron, squashing the mango and heating the fruit to try and heal their knee. When you picture the clothes iron, you will be reminded of the 26th element, Iron.

27. Cobalt

The 26th element is Cobalt. You can break up the name *Cobalt* into "*co*" and "*bolt*". Imagine the clothes iron the doctor is using is very old and broken down. It needs to be held together with two large co-joined bolts or co-bolts. When you picture these bolts, imagine they're like the bolts you see in pictures of Frankenstein. Frankenstein has bolts stuck in either side of his neck. Picture similar co-bolts stuck into each side of the clothes iron holding it together. Make them extra-large and colorful so they stand out. "Co-bolts" will remind you of the 27th element, Cobalt.

28. Nickel

The 28th element is Nickel. A nickel is also a *5 cent coin*. Imagine the injured Vikings have been healed and to prove their fitness, they need to pick up a heavy bag full of nickel coins. They need to pick it up and then drop it overboard. It was a curious medical treatment. The Vikings struggle to pick up the huge heavy bag of nickels and drop it overboard. Imagine the sound of the nickels against each other in the bag, and the sound as the bag hits the water. The important thing is to remember the nickels, as they will help you recall the 28th element, Nickel.

29. Copper

The 29th element is Copper. Once the Vikings drop the bag of nickels overboard, they then have to dive underwater to retrieve it. They use long *copper pipes* as breathing tubes so they can go deep underwater. Picture the bizarre image of each Viking jumping into the water holding onto a long red copper pipe. Imagine them breathing through the pipe, mouth stuck on one end and sucking hard to get enough air to breath. When you think of this rudimentary diving equipment you'll also think of the 29th element, Copper.

30. Zinc

The 30th element is Zinc. *Zinc* sounds like sink, a *kitchen sink*. Visualize one of the Vikings walking around underwater with his copper pipe breathing tube and looking for coins. He kicks his toe on something hard and he sees it's a sink made of zinc. Imagining him walking around and clunk, he hurts his toe and looks down. He sees the huge old sink made of zinc, and that reminds you of Zinc, the 30th element.

Summary

Okay, that's the next 10 so it's time again for a quick revision.

21. Beginning from 21, we have scandium. You remember it by Scandinavia and the Vikings. The Vikings are the gladiators in the mini-Coliseum, with blonde hair and horned helmets. Vikings remind you of Scandinavia and *Scandium*.

22. After winning their battles, they head back to their ship, the Titanic, reminding you of *Titanium*.

23. Because the Titanic is so large, there is a van

aid (aid van) driving around on the deck treating all the wounded Vikings. Van aid equals *Vanadium*.

24. At each stop, a doctor leaps out of the van aid, and he is made completely of chrome. That prompts you to remember *Chromium*.

25. The first treatment he gives each Viking is a mango to put on their knees. Mango knees gives you *Manganese*.

26. The chrome doctor then uses a heavy clothes iron to squash and heat the mango on their knees. The clothes iron reminds you of *Iron*.

27. The iron itself is old and broken and needs to be held together with two large co-joined bolts. These co-bolts remind you of *Cobalt*.

28. After the treatment, the Vikings lift up a huge bag of nickels and drop it overboard. The bag of nickels reminds you of the 28th element, *Nickel*.

29. The Vikings need to jump overboard with a long copper pipe to use as breathing tube to dive and retrieve the nickel coins. Copper pipes make you recall *Copper*.

30. One of the Vikings walking underwater stubs his toe on a sink made of zinc. The sink gives you the 30th element, *Zinc*.

6 GALLIUM to ZIRCONIUM (31-40)

Let's keep going with our memory story. We're up to number 31. Our last image was of a Viking kicking his toe underwater on a kitchen sink.

31. Gallium

Gallium is the 31st element. *Gallium* sounds similar to *galleon*. A galleon is an old style of ship made from timber. Imagine the Viking who'd stubbed his toe manages to drag the kitchen sink back up to the surface. When he gets there, he looks around to find the Titanic. To his surprise, it has changed into an old galleon ship. Picture it as an old-style pirate ship, and when you think of that galleon, you'll remember the 31st element, Gallium.

32. Germanium

The 32nd element is Germanium. *Germanium* was named after *Germany* and that's how you'll remember it. Visualize the galleon ship and picture *Adolf Hitler* walking out onto the deck. Adolf Hitler will represent Germany. Picture what Adolf Hitler would look like standing on the deck of this old-style pirate ship. As you visualize Adolf Hitler, you'll associate him with Germany and the 32nd element, Germanium.

33. Arsenic

Arsenic is the 33rd element. Imagine Adolf Hitler standing on the deck of the ship when a big letter R walks up with a knife and puts a little snick in his leg. A snick is a small notch or cut, so it's an *R snick*. The knife has sliced through Hitler's pants and put a snick in his leg, and a little blood starts to trickle down. When you picture the letter R standing next to Adolf Hitler with a snick in his leg, you'll realize it's an R snick and remember the 33rd element, Arsenic.

34. Selenium

The 34th element is Selenium. *Selenium* we can associate with *Celine Dion*, the singer. If you don't know who Celine Dion is, picture anyone you know named Celine or Selena. Imagine Celine Dion appears on deck next to Adolf Hitler. She gives the snick on his leg a little pinch and then starts singing to him. Celine is wearing a long flowing gown, straight of the stage at Las Vegas, and starts singing in her powerful voice to Adolf Hitler. When you visualize and remember Celine Dion, you will remember the 34th element, Selenium.

35. Bromine

The 35th element is Bromine. When you look at *"Bromine"*, it can be split into *"bro"* and *"mine"*. In the middle of Celine Dion's song, a trapdoor suddenly opens up beneath her. She falls through the trapdoor and lands with a thud in a mine. There are miners digging with shovels all around Celine, and they look at her strangely in her sparkling gown. Imagine all of the thousands of miners all look like your brother or your *"bro"*. If you don't have a brother, imagine someone else's who you know quite well. The mine is filled with thousands of copies of your brother, making it a "bro mine". Picture all of your bro's digging in the "bro mine" and you'll associate it with the 35th element, Bromine.

36. Krypton

The 36th element is Krypton. *Kryptonite* is the substance that makes Superman lose his superpowers, so let's use that association. Picture again all the miners who look like your brother. Imagine they're digging away and uncover some kryptonite. It is glowing green and causes Superman to suddenly fall out of the sky because his super powers have been taken away. Visualize that glowing green kryptonite, Superman falling from the sky, and you'll remember the 36th element, Krypton.

37. Rubidium

Rubidium is the 37th element on the periodic table. When you look at the name *Rubidium*, the first part of it looks like "*ruby*". Imagine Superman lands on his head with a bump and when he gets up he's looking groggy. There is now a ruby embedded in his forehead. Visualize that shiny, glowing, red gemstone in Superman's head and you will remember "ruby" and the 37th element, Rubidium.

38. Strontium

The 38th element is Strontium. *Strontium* sounds a bit like "*strong tea*". A bit, ok? Picture Superman again, standing very groggily with the ruby stuck in his head. The bro miners are try to revive him. They give him a cup of strong tea to help him out. Imagine Superman drinking the intensely strong tea. It's dark like midnight because it was brewed so long. Superman says, "Oh, that's a really strong tea" and it stains his tongue black. When you remember strong tea, you will remember the 38th element, Strontium.

39. Yttrium

The 39th element is Yttrium. "*Hit tree*" sounds like the start of "*Yttrium*". Imagine Superman as he finishes his tea, and he starts to fly away again but he's not feeling 100%. He doesn't fly straight and accidentally runs into a tree. He hits a tree. When you visualize Superman flying straight into a tree and knocking himself out again, you will remember yttrium. "Hit tree" is the 39th element, Yttrium.

40. Zirconium

The last element in this section is Zirconium, the 40th. The beginning of the name *"Zirconium"* sounds similar to *"fir cones"*. Let's imagine the tree Superman hit is a fir tree or pine tree. When he hits it, thousands of fir cones fall to the ground and blanket the area around the tree. You can't see any grass or soil because it's been completely covered with thousands and thousands of fir cones. Picture those "fir cones" and you'll associate them with the 40th element, Zirconium.

Summary

It's time to revise the elements from 31 to 40.

31. For Gallium we remember the galleon ship. The Viking dragged the kitchen sink up to the surface and found the Titanic had changed into an old galleon style pirate ship. Galleon equals *Gallium*.

32. Adolf Hitler appears on the deck of the galleon. As Adolf Hitler represents Germany you are reminded of *Germanium*.

33. The letter R walks up to Adolf Hitler, talks out a knife and with a quick slice, puts a little snick into Hitler's leg. It's an R snick, linking to the element *Arsenic*.

34. Celine Dion appears and gives the snick on Hitler's leg a quick pinch, and starts singing to him at the top of her voice. Celine reminds you of *Selenium*.

35. In the middle of a song, Celine falls through a trapdoor and lands in a mine of digging workers. The thousands of workers all look like your brother – it's a "bro mine". You remember *Bromine*.

36. The miners uncover some kryptonite, causing Superman to fall out of the sky. Kryptonite reminds you of *Krypton*.

37. Superman lands on his head, embedding a ruby into his forehead. Ruby is associated with the element *Rubidium*.

38. In an effort to revive Superman, the miners give him an intensely strong tea, and you remember S*trontium*.

39. Superman drinks the tea and flies away before he's fully recovered. As a result, he hits a tree. Hit tree equals *Yttrium*.

40. The tree is a fir tree and thousands of fir cones fall to the ground. Fir cone reminds you of *Zirconium*, the 40th element of the periodic table.

7 NIOBIUM to TIN (41-50)

Onwards with our story. When we left off you were visualizing fir cones blanketing the ground around the tree.

41. Niobium

We need to link that image with Niobium, the 41st element. Look at the name and break it up. The first half of it looks similar to *"knee oboe"*. Imagine a giant squirrel is attracted by all of the fir cones and he absolutely loves fir cones. He is so happy to discover this abundance of fir cones he starts to play his knee oboe. It's an oboe played with the knees. Picture the giant squirrel squeezing music out of it between his knees. Once you can see the squirrel playing a knee oboe, you will remember the 41st element, Niobium.

42. Molybdenum

The 42nd element is Molybdenum. It's another challenging name but if you use a bit of imagination you can turn it into the words "*mole bee denim*". So, we have our squirrel playing his knee oboe. Imagine the squirrel is wearing denim jeans and because the knee oboe is rubbing against his jeans, he wears a hole straight through them. You can now see a large mole on his knee. Suddenly a bee flies down and stings him on the mole. We have a mole, a bee and denim. You have a few bits to visualize there. The knee oboe has worn a hole in the squirrel's denim jeans and a bee stings him on the mole on his knee – mole bee denim. Again, it's a bit of a stretch and it's a bit ridiculous but if you create a really strong visual image, you'll remember it for sure. "Mole bee denim" reminds will remind you of the 42nd element, Molybdenum.

43. Technetium

The 43rd element Technetium. This name will be associated with "*tech net*" or technical net. Imagine the squirrel is very angry at the bee which stung him on the knee. The squirrel chases after the bee with his technical net. His tech net is like a butterfly net with flashing lights on it. Maybe it's even remote controlled – it's a very technical net. "Tech net" will help you recall the 43rd element, Technetium.

44. Ruthenium

The 44th element is *Ruthenium*. If you look at that word, the first part of it can be broken into "*rut hen*". Imagine the squirrel is chasing the bee with a technical net and he suddenly trips on a rut or a small ditch. A rut is a small ditch or hole. The squirrel trips on the rut and falls and lands on a hen. The hen is naturally very surprised and quite upset. Imagine the squirrel running and tripping on the rut and falling onto a hen. Picture first the rut and then the hen, and you'll remember the 44th element, Ruthenium.

45. Rhodium

Next is Rhodium, the 45th element. *Rhodium* sounds similar to *rodeo*. Visualize the startled and surprised hen, leaping up and jumping around. The squirrel is clinging onto the back of the hen for dear life. The hen is bucking around trying to get the squirrel off its back, like a rodeo. We're not talking about a cowboy on a bull; we're talking about a squirrel on a hen – that kind of rodeo. The rodeo will remind you of the 45th element, Rhodium.

46. Palladium

The 46th element is Palladium. The first part of *Palladium* sounds like "*pa laid*". With our rodeo image of the hen and the squirrel, imagine bizarrely, the hen suddenly morphs into your pa – your father – and lays an egg. Again, it's a very ridiculous image. Visualize it strongly and you'll never forget it. The hen changes into your pa, and lays an egg. So "pa laid" indelibly reminds you of the 46th element palladium.

47. Silver

The 47th element is Silver. Clearly picture the egg your pa laid. It rolls onto the ground and cracks open. As the shell cracks open, a glittering *bar of silver* is revealed inside it. It's surprising an egg laid by your pa should have anything but a yolk inside. Instead, it's a shiny bar of silver. That will remind you of the 47th element, Silver.

48. Cadmium

The 48th element is Cadmium. *Cadmium* sounds like caddie, as in a *golf caddie.* Imagine a golf caddie – the guy who carries the bag of golf clubs for a player – appears, picks up the silver, pulls the old eggshell off it and puts the silver bar in his bag of golf clubs. You have no idea why he did that, but when you think of the caddie, you'll also think of the 48th element, Cadmium.

49. Indium

The 49th element is Indium. *Indium* sounds like *Indian*, someone from India. Visualize the caddie starting to walk away with his golf club bag and the silver bar inside it, and suddenly a large curry thuds into the side of the bag. It's not in a pot, it's just a big and hot beef curry and it splatters all over the golf bag. The caddie is startled and turns around.

There in front of him is a ferocious looking Indian lady wearing a traditional Indian dress or sari. The sari is blindingly bright in all the colors of the rainbow, and threaded with gold. Make the sari flash in the sunlight as the Indian lady moves. She looks incredibly angry and the red bindi on her forehead seems to pulse and glow with her mood. When you see that angry Indian lady in your mind, you'll recall the 49th element, Indium.

50. Tin

The 50th element is Tin. The naturally association with *Tin* is a *tin can*. Imagine the caddie, quite frightened after being attacked with food by the angry Indian. In retaliation he throws a tin can of food at the Indian. Visualize the scared caddie as he pulls out a can of spaghetti and in his fear, he hurls the tin can at the Indian. When you think of that flying tin can, think of the 50th element, Tin.

Summary

And now you've memorized the elements up to 50. Let's revise to make sure all those weird images are securely in your mind.

41. Niobium. We had the thousands of fir cones lying around on the ground which were discovered by a giant squirrel. In the squirrel's joy, he starts to play his knee oboe. He squeezes music out of the oboe with his knees. Knee oboe gives *Niobium*.

42. While he is playing the knee oboe, the squirrel wears a hole in his denim jeans on the knees. A mole is revealed, which is stung by a bee. The mole stung by a bee through the denim is shortened in your mind to "mole bee denim", reminding you of *Molybdenum*.

43. The squirrel is very angry at the bee and chases after it with his technical net which has flashing lights on it. It's a technical net or a "tech net" – *Technetium*.

44. While the squirrel is chasing after the bee with his net, he trips in a rut and falls and lands on a hen. "Rut hen" is associated in your mind with *Ruthenium*.

45. The startled hen leaps up and starts bucking around with the squirrel on its back like a rodeo. Rodeo reminds you of *Rhodium*.

46. Quite bizarrely, the hen changes into your pa (your father) who lays an egg. "Pa laid" an egg – *Palladium*.

47. The egg itself rolls onto the ground and cracks open, revealing a glittering bar of silver. It's valuable and quite surprising in an egg. Not so surprising is that a bar of silver reminds you of *Silver*.

48. A golf caddie appears and puts the silver bar in his golf bag. The caddie reminds you of *Cadmium*.

49. As he walks off with the silver bar in his bag, a curry thuds into the side of the golf bag, thrown by an Indian lady wearing a blindingly bright sari. Indian reminds you of *Indium*.

50. The scared caddie throws a tin can of food at the Indian to try and scare her away, and the flying tin can helps you remember *Tin*. You're now up to number 50 and traveling really well!

8   ANTIMONY to NEODYMIUM (51-60)

We are now up to number 51 on the periodic table, and the scared caddy had just thrown a tin can of food at the Indian.

51. Antimony

The 51st element is Antimony. When you look at "*Antimony*" you are going to think of the words "*ant money*". A strange concept, I know. The Indian lady catches the tin can which was thrown at her, and sells it to an army of ants. There are millions of these ants and they love tin cans of food. In exchange, the ants pay the Indian with a pile of extremely small dollar bills. They're small because it's ant money, absolutely tiny. Any money will remind you of the 51st element, Antimony.

52. Tellurium

The 52nd element is tellurium. When you look at the

word *Tellurium,* I want you to see a teller, like a *bank teller* who works at the bank, giving and receiving money. Imagine the Indian takes the pile of very small ant money to the bank, but the teller behind the desk won't accept it because it's too small. It's not real money, it is ant money. Visualize the teller being quite indignant and saying, "No, we don't accept ant money here. We're a real bank." When you think of that indignant teller, you will recall the 52nd element, Tellurium.

53. Iodine

Iodine is the 53rd element on the periodic table. *Iodine* sounds similar to the sentence "*I owe Dean*". Now imagine the Indian, who just had her money rejected, starts to get quite upset and she screams, "But, I owe Dean. I owe Dean money." She doesn't know James Dean is dead. She owes James Dean's money but she doesn't realize the debt is no longer valid because James Dean's is dead but she is quite upset all the same. Picture her screaming "I owe Dean" and you'll be reminded of the 53rd element, Iodine.

54. Xenon

The 54th element is Xenon. When you pronounce *Xenon,* the X sounds like Z, so the word sounds like "*Zen on*". Imagine the Indian lady is upset and screaming, "I owe Dean" and a Zen master appears and tries to calm the Indian down. He does that by sitting cross-legged on the Indian. It's a classical

Zen pose with his crossed legs and crossed arms. He's sitting on the Indian. We have the Zen on the Indian. "Zen on" the Indian is Xenon. That's how you remember the 54th element, Xenon.

55. Cesium

Cesium is the 55th element. *Cesium* is similar to *caesarean.* Caesarean is the medical procedure for delivering a baby where the doctors or the surgeons make an incision in the lower belly of the mother and remove the baby through the incision. Imagine the Zen master who is sitting on top of the Indian suddenly performs a caesarean on the Indian. Picture the Zen master pulling out a scalpel and makes an incision across the belly of the Indian. He is performing a caesarean which reminds you of the 55th element, Cesium.

56. Barium

The 56th element is Barium. *Barium* sounds like "*bury him*". Naturally, the Indian is furious at the Zen master for attempting this unauthorized caesarean on her and decides to bury him. "Bury him" reminds you of barium. The Indian buries the Zen master so only his head is showing above the ground. The rest of his body, arms and limbs are buried and out of sight. Bury him – the Zen master – reminding you of the 56th element, Barium.

57. Lanthanum

Lanthanum is the 57th element. *Lanthanum* sounds

a bit like "*lantern*". Picture the Zen master who is buried except for his head sticking out above the ground. Now his head begins to shine like a lantern. It glows a bright yellow and starts to flash on and off too. Lantern will help you recall the 57th element, Lanthanum.

58. Cerium

Cerium is the 58th element. *Cerium* is pronounced a bit like *cereal*. As the Zen master's head starts to shine like a lantern, it begins to rain breakfast cereal on top of his exposed head. There's Fruit Loops, Cheerios, Corn Flakes and Granola. It's all raining down on top of his glowing lantern head. Picture the cereal raining down and you will remember the 58th element, Cerium.

59. Praseodymium

The 59th element is Praseodymium. This is going to be a tough one but let's give it a try. Break it up and the beginning of it sounds like "*praise o dim*". Imagine the little Fruit Loops and Cheerios are animated. They've got arms and legs and they start to worship the Lantern head of the Zen master, crying out "Praise o dim light. Praise o dim." "Praise o dim" will remind you of praseodymium. Just imagine all of the little Fruit Loop and Cheerio worshippers yelling out, "Praise o dim" at the lantern head and you will remember the 59th element, Praseodymium.

60. Neodymium

The 60th element is Neodymium. We're going to stretch for this one too. If you take the first part of "*Neodymium*" you can turn it into "*neo dim*". In the movie 'The Matrix', Keanu Reeves' character was named Neo. Imagine Neo from The Matrix suddenly appears in our visual memory story, but walks straight into a tree. All of the Fruit Loops and Cheerios watch him walk straight into the tree, and start laughing and crying out, "Neo dim. Neo dim." They're not impressed with the intelligence level of Neo from The Matrix so, "Neo dim" will remind you of Neodymium, the 60th element.

Summary

Let's revise the story from 51 to 60.

51. The 51st element is antimony and your mental link is ant money. The Indian lady sells the tin can to the ants in exchange for a pile of ant money, which reminds you of *Antimony*.

52. The Indian takes the ant money to the bank but the bank teller indignantly tells her to go away. Teller links to *Tellurium*.

53. The Indian gets upset and starts screaming "I owe Dean". She owes James Dean money. And *Iodine* is the next element.

54. A Zen master appears and tries to calm the

Indian lady, but he does it by sitting on top of her. "Zen on" the Indian gives you *Xenon*.

55. The Zen master then attempts to perform a caesarean on the Indian, which makes you remember *Cesium*.

56. In retaliation, the Indian lady buries the Zen master so only his head is above ground. "Bury him" relates to *Barium*.

57. The Zen master's head is the only thing showing above the ground and it starts to shine like a lantern. Lantern reminds you of *Lanthanum*.

58. Breakfast cereal of all kinds begins to rain down on top of the Zen master's lantern head, almost burying it. Cereal is associated with *Cerium*.

59. The little Fruit Loops and Cheerios become animated and begin to worship the lantern head yelling out "Praise o dim. Praise o dim light." "Praise o dim" makes you think of the 59th element, *Praseodymium*.

60. Then, much to the amusement of the cereal, Neo from The Matrix appears and walks straight into a tree. The cereal, all the Fruit Loops and Cheerios start to cry out, "Neo dim." Neo dim gives you the final element in this section, *Neodymium*.

9 PROMETHIUM to YTTERBIUM (61-70)

61. Promethium

On to the 61st element which is *Promethium*. What are we going to do here? Let's go with "*prom Ethel*". Ethel is an old fashioned name and it suits her. She's on the way to her prom, but she's wearing an old fashioned dress, like something from 'Gone with the Wind'.

Neo has just gone smack into a tree and the cereals are crying "Neo dim." Then prom-Ethel rushes up to help Neo stand up again. Ethel is dressed in her old-fashioned prom dress and she's struggling to help Neo stand up because Neo is a big guy and she's just a teenager wearing a large puffy dress. Picture the dowdy "prom-Ethel" and you'll recall the 61st element, Promethium.

62. Samarium

The 62nd element is Samarium. *Samarium* sounds similar to *Samaritan*. Prom-Ethel is struggling to help Neo stand up when a Good Samaritan rushes up and helps Ethel get Neo standing up. The Good Samaritan is straight out of the Bible and he's wearing long sweeping robes and he's kind and gentle like a Good Samaritan should be. Picture that very kind and gentle Good Samaritan and it will remind you of the 62nd element, Samarium.

63. Europium

Europium is the 63rd element. *Europium* which is named after *Europe*. Imagine the Good Samaritan is incredibly kind and gives Ethel a trip to Europe, the land of the Eiffel Tower, the Swiss Alps and the Colosseum. The Samaritan gives Ethel a plane ticket to go to Europe. Picture Ethel daydreaming about all of those iconic landmarks in Europe, and you will associate that image with the 63rd element, Europium.

64. Gadolinium

Next is the 64th element, *Gadolinium*. Number sixty-four is another challenging one. If you're a bit creative and break that up into a couple of words, you can go with "*gay doll*". Imagine Ethel is on the plane traveling across to Europe and she's playing with a very happy or gay doll. Visualize it as being a doll that has a variety of different smiles and makes different kinds of laughing and giggling noises, because it is so incredibly gay. Focus on that gay doll and you'll be able to remember the 64th element, Gadolinium.

65. Terbium

The 65th element is Terbium. *Terbium* sounds like *turban*. Imagine Ethel on the plane playing with her gay doll. The rest of the plane is completely full of men wearing turbans. Focus on the turbans. Make them really high turbans which rub the ceiling of the plane and block Ethel's view of the large TV up front. Clearly picture all of those turbans and be reminded of the 65th element, Terbium.

66. Dysprosium

The 66th element is *Dysprosium*, another challenging one. Let's go with "*dissing prose*". The men wearing turbans are all quite talkative and speak in very eloquent prose. Prose is written or spoken language that isn't structured like poetry. Imagine the turban wearing men talking in eloquent prose when an argument breaks out and they begin disrespecting or dissing, the way each other are speaking. They are dissing prose. Visualize these men getting angry as they 'diss' prose, and you'll remember the 66th element, Dysprosium.

67. Holmium

Next is Holmium the 67th element, which sounds like "hole me in". Imagine you've been watching this visual story movie, but you've been watching it from inside the plane. You're in the plane with Prom-Ethel and the turbaned and irate men. Suddenly, you fall into a hole and free fall for a few seconds. From your perspective, it's 'me in the hole' or "hole me in" if you twist it around. Feel the sudden dropping sensation as you go into the hole. "Hole me in" reminds you of the 67th element, Holmium.

68. Erbium

The 68th element is Erbium. *Erbium* reminds me of *herbs*. After falling in the hole, you make a soft landing on herbs. There are huge bushes of basil, thyme and rosemary. Imagine how fragrant they smell. You get to your feet and walk all over the herbs because they completely cover the ground. Picture the fragrant herbs and remember the 68th element, Erbium.

69. Thulium

Thulium is the 69th element. Let's say *Thulium* is pronounced as "*tool-ium*" and try to work a *tool* into the story. As you're walking on the fresh herbs you stand on something sharp which pierces your foot. You look down and realize you've stood on a tool, a sharp rake. Visualize yourself standing on that sharp tool and hurting your foot, and you'll be reminded of the 69th element, Thulium.

70. Ytterbium

The 70th element is Ytterbium. *Ytterbium* sounds like "*hit herbs*". Imagine you pick up the tool you've stood on and use it to hit herbs. You flatten a large patch of herbs by hitting them with the tool. Maybe it's good stress release. Picture yourself hitting herbs and you'll remember the 70th element, Ytterbium.

Summary

It is revision time again.

61. Remember prom-Ethel helping Neo stand up? Picture her and you'll remember *Promethium*.

62. And the kind and gentle Good Samaritan who rushes to her assistance reminds you of *Samarium*.

63. He's so kind he gives prom-Ethel a plane ticket to Europe, home of the Eiffel Tower and other historical and tourist icons, which links to *Europium*.

64. On the plane to Europe prom-Ethel plays with her happy and gay doll, associated in your mind with *Gadolinium*.

65. Watching her play with the gay doll is a plane full of men wearing high turbans. The turbans remind you of *Terbium*.

66. The turban wearing men start to get agitated and begin to 'diss' prose (disrespect each other's prose), and that sounds like *Dysprosium*.

67. From your vantage point of this story you suddenly fall into a hole – hole me in – *Holmium*.

68. When you land at the bottom of the hole you find yourself on a garden full of fresh herbs. You remember *Erbium*.

69. Standing up, you then walk around but stand on a tool and hurt your foot. The tool reminds you of the slightly mispronounced *Thulium*.

70. Taking the tool in your hands you use it to hit the herbs and flatten them. "Hit herbs" is linked in your mind to *Yytterbium*.

10 LUTETIUM to MERCURY (71-80)

71. Lutetium

Here we are at Lutetium, the 71st element on the periodic table. Looking at *Lutetium*, you could break up the first part into *"lute eat"*. A lute is an old-fashioned string instrument similar to a mandolin with a body shape rounder than a guitar.

Having flattened the herbs all around you, a lute suddenly stands up. It's right in the middle of the flat herbs, and it starts to rip up the flattened herbs and eat them. Imagine the lute eating the herbs, and this lute can really eat. It is snuffling around on the ground, gobbling herbs. "Lute eat" will remind you of the 71st element, Lutetium.

72. Hafnium

The 72nd element is Hafnium. *Hafnium* is a challenging name, but let's go with *"have no arm"*.

Imagine "have no arm" being said with an accent, maybe Irish or German, whatever accent will get you from "have no arm" to hafnium. Visualize the lute standing up and screaming with frustration, "I have no arm. Hafnium." The lute is shaking its arm stump in frustration. "Have no arm" becomes the 72nd element, Hafnium.

73. Tantalum

Tantalum is the 73rd element. We'll associate *Tantalum* with *"tan tail"*. Tan is a color similar to coffee. When you picture a tan tail in your mind, make it like a fox tail without a body. It's just a tail and it is colored tan. While the lute is standing in the middle of the herbs complaining about having no arm, the fluffy tan tail rustles through the herbs and brushes past the lute. Make the lute sneeze or laugh when the tail touches it. Clearly picture that tan tail scurrying by itself past the lute, and you'll have a link to the 73rd element, Tantalum.

74. Tungsten

Next is tungsten, the 74th element. *Tungsten* sounds very close to *"tongue stun"*, which will work well. Imagine you're standing on the edge of the herb garden where you've been watching the lute and tan tail. Y ou continue watching as the tan tail crawls past the lute, up your body, into your mouth and onto your tongue. Your tongue is completely stunned. There is a fluffy, furry tail on your tongue. Your tongue is stunned and surprised at being covered with fur. You are "tongue stun" and reminded of the 74th element, Tungsten.

75. Rhenium

Rhenium is the 75th element. To remember *Rhenium* you'll picture a small *wren*, a little type of bird. Visualize a small bright blue colored wren as it swoops down out of the sky and plucks the tan tail out of your mouth. Imagine the little blue wren swooping down, attacking your mouth and grabbing the tan tail in its beak and pulling it out. The wren will remind you of the 75th element, Rhenium.

76. Osmium

The 76th element is Osmium. After the wren has grabbed the tail out of your mouth, it asks you "Are you the Wizard of Oz?" Now you can talk again without the tan tail stuffed in your mouth you reply a surprised voice, "*Oz? Me?*" When you picture yourself saying "Oz? Me?" as you stammer, trying to imagine yourself in the movie of The Wizard of Oz, dressed as one of the characters or as Oz himself. "Oz? Me?" will link in your memory to the 76th element, Osmium.

77. Iridium

Iridium is the 77th element. If you look at that name, *Iridium* the first part is the words, "*I rid*". While you're imagining yourself as the Wizard of Oz, an evangelical preacher appears. He's a traditional preacher, wearing a flowing gown and big cross around his neck. He's afraid you're filled with demons and evil and performs an exorcism on you. He places his hand on your forehead and yells "I rid you!" and pushes you backwards. When you see him in your mind yelling "I rid you", you'll remember the 77th element, Iridium.

78. Platinum

Platinum is the 78th element. *Platinum* sounds like "*plate numb*". The preacher pushes you backwards and you fall on the ground. As you land, you land on a plate. It's a talking plate consumed with pain from you on landing on him, and is no longer able to speak. He is numb and feels no pain. He can't feel anything because you've crushed him. "Plate numb" reminds you of the 78th element, Platinum.

79. Gold

The 79th element is Gold. The talking plate has arms and legs, but is now lying numb on the ground, spread-eagled and trying to recover. Suddenly, a huge *boulder of gold* falls out of the sky and lands on him. He's taking a beating. First it was you, now this enormous gold nugget. Picture that shiny gold boulder sitting on the numb plate, and you'll easily remember the 79th element, Gold.

80. Mercury

The 80th element is Mercury. If you've ever seen liquid mercury, you'll know it's like a thick liquid and silvery in color. Imagine the gold boulder sitting on the plate for a few seconds and then it turns into liquid mercury in an instant. Since it was such an enormous boulder, it creates a flood of liquid mercury. Imagine this flood of thick silvery liquid mercury. It's a torrent, flooding the area. When you remember that flood of liquid mercury you'll be reminded naturally of the 80th element, Mercury.

Summary

And now you've completed up to the 80th element. It's time again for a quick revision.

71. Seventy-one was lutetium. A lute stands up out of the herbs and begins to eat them. "Lute eat" links to *Lutetium.*

72. The lute screams in frustration, "I have no arm, have no arm." This is said in an accent and sounds like *Hafnium.*

73. A tan tail without a body rustles past the lute. It's covered in fluffy, tan colored fur. Tan tail reminds you of *Tantalum.*

74. The tan tail crawls over to where you're standing, crawls up your body, into your mouth and onto your tongue. You are tongue stunned. "Tongue stun" gives you *Tungsten,* the 74th element.

75. A small bright blue wren swoops down and

grabs the tan tail out of your mouth. The wren will remind you of *Rhenium*.

76. After grabbing the tail out of your mouth, the wren asks if you are the Wizard of Oz. You cry in surprise "Oz? Me?" This prompts you to recall *Osmium*.

77. Then an evangelical preacher appears. He performs an exorcism on you, pushing you backwards on the forehead and yells "I rid you!" "I rid" becomes *Iridium*.

78. You fall backwards and land on a talking plate. The plate is numb with pain from you landing on him. He's a numb plate, linking to *Platinum*.

79. While the numb plate is lying on the ground trying to catch his breath, a large boulder of gold falls out of the sky and lands on him. The huge gold nugget reminds you of *Gold*.

80. In an instant, the enormous gold boulder turns into liquid mercury and creates a thick silvery flood of liquid mercury. You use that image to remember *Mercury*.

You have achieved something now – you have memorized elements 71 to 80.

11 THALLIUM to THORIUM (81-90)

81. Thallium

The 81st element on the periodic table is Thallium. *Thallium* rhymes with *valium*. Valium is a common drug which makes people relaxed and spaced out.

Imagine floating on the surface of the flood of liquid mercury are hundreds of spaced out people. They're spaced out because they're on valium. Picture hundreds of people floating by, all drugged and relaxed because they're on valium. The people on valium remind you of the 81st element, Thallium.

82. Lead

The 82nd element is Lead. Pencils are sometimes called *lead pencils*, as if they used a combination of lead and carbon. Visualize all of the people on valium, and as they float by they take out lead pencils and use them like blow guns. They blow on one end of the pencil and make the lead shoot out the other end. They're trying to shoot each other with their lead pencils as they float along. Picture them using the lead pencils like blow guns and you'll remember Lead, the 82nd element.

83. Bismuth

The 83rd element is Bismuth. What will we do here? Let's go with *buzz moth*. Imagine one of the pieces of lead hits a tiny flying moth. There's an immediate strange reaction and the moth grows to the size of a car and starts buzzing loudly. It's a buzz mouth, and reminds you of the 83rd element, Bismuth.

84. Polonium

Polonium is the 84th element. *Polo* is the obvious image to use. Picture the large buzz moth flying over a game of polo. There are polo ponies being ridden all over the field in pursuit of a small white ball. The buzz moth attacks one of the polo players on the back of a polo pony. Clearly visualize that game of polo and you'll be reminded of the 84th element, Polonium.

85. Astatine

Astatine is the 85th element. What does *Astatine* sound like? We'll go with *ass tattoo*. One of the polo players is being attacked by the moth. The moth is carrying a tattoo needle and gives the polo player an ass tattoo on his arm. It's a tattoo of a donkey or ass. Picture that giant buzz moth giving the polo player an ass tattoo on his arm, and you'll never forget the 85th element, Astatine.

86. Radon

Element 86 is Radon. *Radon* sounds like *ray gun*. Focus on the tattoo of the ass. Imagine the ass or donkey comes alive and starts shooting the polo crowd. The ass isn't using a normal gun, but a ray gun. The spectators dive for cover as laser rays explode around them. Imagine the ray gun and you'll remember the 86th element, Radon.

87. Francium

The 87th element is francium. *Francium* was named after France, so we'll have to work that into the story. The classic image of France is the *Eiffel Tower*. Imagine the ass is firing the ray gun indiscriminately, and one of the rays misses the crowd and hits the Eiffel Tower. Picture the Eiffel Tower as it slowly collapses to the ground with a loud crash. The image of the Eiffel Tower will remind you of France and trigger the name of the 87th element, Francium.

88. Radium

Radium is the 88th element. Your association this time for *Radium* will be "*raiding hymn*". All across Paris, where the Eiffel Tower is located, a deafening raiding hymn starts to sound. It's a loud hymn or song which acts as a warning sign that Paris is being raided and is under attack. They're being attacked by a ray gun toting dolphin. Imagine the streets of Paris echoing with the sound of the raiding hymn. It's an ominous sound warning everyone the city is being raided. "Raiding hymn" reminds you of the 88th element, Radium.

89. Actinium

Actinium is the 89th element. *Actinium* sounds like *acting*. Think of your picture of the streets of Paris echoing with the raiding hymn. The hymn causes the streets to be flooded with thousands of actors. There are mime actors silently acting out scenes, there are movie actors being filmed on camera and there are theater actors quoting Shakespeare. They're all acting. Imagine this scene of mass acting and you'll remember the 89th element, Actinium.

90. Thorium

The 90th element is Thorium. *Thor* is the god of thunder and lightning. Thor is not fond of acting, whether it's mime or anything else. He takes his huge hammer and begins smashing the city of Paris. Picture Thor being upset at the acting and smashing everything in Paris with his hammer, and you'll be reminded of the 90th element, Thorium.

Summary

Alright, that's 81 to 90. Here's a quick summary.

81. You start at thallium which sounds like valium. Hundreds of people drugged up on valium float by on the flood of mercury. Valium reminds you of *Thallium.*

82. All of the Valium drugged people start to shoot each other with lead pencils. They use them like blow guns, shooting each other with the leads. Lead pencils remind you of *Lead.*

83. One of the pieces of lead accidentally hits a moth. It causes a reaction and turns the moth into a giant buzzing moth. Buzz moth links to the element *Bismuth.*

84. The large buzz moth flies over a game of polo and attacks one of the polo players. The polo game reminds you of *Polonium.*

85. The moth attacks the polo player with a tattoo needle and gives him an ass tattoo. Ass tattoo sounds similar to *Astatine.*

86. The tattoo is an ass or donkey. The ass comes alive and begins shooting the polo crowd with a ray gun. Ray gun reminds you of *Radon.*

87. One of the rays from the ray gun misses the crowd and hits the Eiffel Tower, causing it to collapse. The Eiffel Tower links to France and reminds you of the 87th element, *Francium.*

88. All around the city of Paris a raiding hymn begins sounding the warning they're being raided. "Raiding hymn" is associated with *Radium*.

89. Once the raiding hymn starts, thousands of actors come out onto the streets and start acting. There's mime, theatre and movie acting, but it all reminds you of *Actinium*.

90. Thor, the god of thunder and lightning, hates acting and he becomes upset and hits the city with his giant hammer and knocks down all of the actors. Thor will remind you of *Thorium*.

12 PROTACTINIUM to FERMIUM (91-100)

91. Protactinium

Now, we are up to the 91st element – *Protactinium*. Here we go.

After Thor has knocked down all of the actors, there is a single actor left standing. That single actor is miming and pretending to be a pro baseballer. He's a batter but instead of using a bat, he's using a giant tack made of tin. It's a tack made of tin being swung by a pro. Visualize all of those elements – *pro, tack and tin* – and you'll get the 91st element, Protactinium. It's completely ridiculous, making it even stickier in your memory once you've pictured it clearly.

92. Uranium

Uranium is the 92nd element. Imagine the tin tack turning into *radioactive* uranium. It starts to *glow green* and makes the pro base-baller's hands and arms glow green as well. Visualize the tin tack becoming radioactive, glowing green uranium and you will remember the 92nd element, Uranium.

93. Neptunium

Neptunium is next as the 93rd element. The pro baseballer decides it's sensible to get rid of the giant radioactive tin tack. He throws it into the ocean where it is scooped up by *Neptune*. Neptune is the Roman god of water and sea. He just happens to be motoring by, standing on the backs of a couple of giant sharks. Neptune will remind you of the 93rd element, Neptunium.

94. Plutonium

The 94th element is Plutonium. Let's go with *Pluto*, the dog from Disneyland. As Neptune goes sailing by, he's spotted by Pluto the dog, Mickey Mouse's friend. Pluto was walking on the beach but jumps into the water and starts swimming after Neptune. He really wants to catch the intriguing Neptune and his sharks. Remember Pluto the dog and you will remember the 94th element, Plutonium.

95. Americium

The 95th element is Americium. Obviously, it was named after *America*. Picture Pluto swimming after Neptune. Neptune goes out of sight and Pluto swims so hard he hits land back in America (he was on the beach in France). Imagine Pluto swimming under the Golden Gate Bridge and you'll know he's reached America. He swam a long way. When you picture Pluto hitting land in America, you'll remember the 95th element, Americium.

96. Curium

The 96th element is Curium. *Curium* sounds like *curious*. Pluto reaches land and collapses on the American shore, completely exhausted after such a long swim. A very curious cat comes sniffing around him. Remember the saying 'curiosity killed the cat'? It's usually not very sensible for a cat to sniff a dog, so it must be an incredibly curious cat. Picture that extremely inquisitive and curious cat, and you'll be reminded of the 96th element, Curium.

97. Berkelium

Berkelium is the 97th element. It is named *Berkelium* as it was discovered at *UC Berkeley,* a famous college near San Francisco in California. Pluto wakes up and barks loudly at the curious cat sniffing him. The curious cat is frightened for its life, to be barked at by a dog as big as Pluto. The cat runs for its life and hides at UC Berkeley, also known as 'Cal Berkeley'. The college logo is the word 'Cal' in yellow lettering, written in a similar script to Coca Cola, and on a dark blue background. Imagine a giant baseball cap with the 'Cal' logo on it, and associate it with Cal Berkeley. The cat hides underneath that giant cap at Berkeley. Berkeley will remind you of the 97th element, Berkelium.

98. Californium

Californium is the 98th element, named after *California*. Imagine the entire Berkeley College campus begins to shake as an enormous earthquake hits. California is very prone to earthquakes and this is such a large earthquake the entire state of California breaks away and drops into the ocean. Poor California. When you think of California dropping into the ocean, you'll be reminded of the 98th element, Californium.

99. Einsteinium

Albert Einstein is the namesake of the 99th element, *Einsteinium*. Picture a huge figure of Einstein is left standing on the new edge of the West Coast of the United States. California has disappeared and there's a new West Coast. The large Einstein shakes his famous head as he looks down into the water at California, and his wiry grey hair blows in the wind. Picture Einstein standing there and you will be reminded of Einsteinium, the 99th element.

100.   Fermium

The 100th element is Fermium. For *Fermium*, let's go with *'firm'*. Imagine Einstein's entire body suddenly becomes stiff and firm. It firms up and turns Einstein into a giant statue. There's now a giant statue of Einstein standing on the new West Coast. Picture Einstein becoming firm as a statue, and you'll remember the 100th element, Fermium.

Summary

That brings us to summary time, once again.

91. Ninety-one was protactinium. We broke that name into "pro tack tin" and picture an actor pretending to be a pro base-baller using a giant tin tack as a bat. "Pro tack tin" links to the name *Protactinium.*

92. The tin tack turns into radioactive uranium. It glows green in the base-baller's hands, and reminds you of *Uranium.*

93. The radioactive tin tack is thrown into the ocean where Neptune scoops it up. Neptune is sailing by on the back of sharks, and is associated with the element *Neptunium.*

94. Pluto the dog sees Neptune sailing by and leaps into the water and begins chasing him. Picture Pluto the dog and you'll remember *Plutonium.*

95. Pluto swims so hard he reaches America, and you remember *Americium.*

96. The exhausted Pluto collapses on the American shore where he is sniffed at by a very curious cat. The curious cat prompts you to remember *Curium.*

97. When Pluto wakes up and barks at the cat, the frightened cat runs and hides at UC Berkeley under a giant Cal Berkeley cap. Berkeley links to *Berkelium.*

98. Suddenly, the entire campus starts to shake as an enormous earthquake hits. It's so big the entire state of California drops into the ocean. As you picture California breaking away, you're reminded of *Californium*, the 98th element.

99. Albert Einstein is standing shaking his famous head on the new edge of the West Coast, looking at where California used to be. The image of Einstein reminds you of *Einsteinium*.

100. Then, Albert Einstein's body became very firm and turns into a giant statue. Because it is so firm, you are reminded of *Fermium*.

And that's the first 100 elements.

13 MENDELEVIUM to COPERNICIUM (101-112)

We are on to the final section, where you'll learn 101 to 112. The elements go up to 118 in most periodic tables, but from 113 to 118 they only have temporary names which we won't memorize.

101. Mendelevium

Mendelevium is the 101st element. For *Mendelevium* you can use "*mental leave*" to prompt your memory. After Einstein's entire giant body firms up into a giant statue, men in white coats come and carry Einstein away to a mental institution. He's put on "mental leave". Picture Einstein, the towering genius, having to go on "mental leave" and that will remind you of the 101st element, Mendelevium.

102. Nobelium

Next is Nobelium, the 102nd element. Picture Einstein in a sparsely furnished room in the mental institution. He's lying on his bed when he blinks and comes back to reality. His body is no longer firm like a statue. He's wondering where he is and presses the call button for a nurse. Despite urgently pressing the button there is *no bell*. Imagine Einstein frantically pressing the button but being met with silence because there is no bell. "No bell" will remind you of the 102nd element, Nobelium.

103. Lawrencium

The 103rd element is Lawrencium. It was obviously named after Lawrence – whichever Lawrence that may be. We will picture *Lawrence of Arabia* in his long Arabian clothes, long flowing gown and head cover. After Einstein presses his call button but there is no bell, only silence, Lawrence of Arabia walks into his room. Visualize Lawrence of Arabia walking in with his long white flowing gown. Einstein is quite confused because he was expecting a nurse. Picture Lawrence of Arabia and link that image to the 103rd element,Lawrencium.

104. Rutherfordium

The 104th element is Rutherfordium. Lawrence of Arabia realizes he's in the wrong room. He didn't come here to visit Albert Einstein. He goes into the next room where his brother Ford is. *Brother Ford* will remind you of *Rutherfordium*. Lawrence of

Arabia's brother Ford has a very large blue Ford emblem tattooed on his forehead. Ford is his name and he's decided to brand himself as a Ford as well. Your mental image of "Brother Ford" reminds you of the 104th element, Rutherfordium.

105.   Dubnium

Number 105 is Dubnium. For *Dubnium*, we'll use double knee or "*dub knee*" for short. You've pictured Brother Ford with the Ford tattoo lying in bed. Now he gets out of bed. When he stands up, you can see his legs bend very strangely because he has a double knee. He has a knee at the front and the back of each leg and they bend both ways. He walks in a very bouncing manner because his legs bend both ways because he has double knees. Double knee or "dub knee" will remind us of the 105th element, Dubnium.

106.   Seaborgium

The 106th element is Seaborgium. *Seaborgium*, named after Seaborg. Let's go with "*C board*", a special type of surfboard. It's shaped like a letter C, and designed especially for Brother Ford and his double knees. Once Brother Ford gets out of bed, he grabs his C board and heads to the pool. Imagine what that C-shaped board looks like, and be reminded of the 106th element, Seaborgium.

107.   Bohrium

Element 107 is Bohrium, named after the scientist

Niels Bohr. Bohr sounds like "*bore*". Brother Ford reaches the pool with his C board and begins talking to people. Unfortunately he's such a bore – a boring person – people fall asleep listening to him. They fall asleep and collapse into the pool and drown. Picture all those people falling asleep and into the pool, and the surface of the pool is covered with floating bodies. Brother Ford is a complete bore. Bore links in your memory to the 107th element, Bohrium.

108.   Hassium

The 108th element is Hassium. "*Haze*" will remind you of that name. Picture again the scene of the pool covered in floating bodies, and watch as a haze or a white fog slowly descends over the pool. You can no longer see any of the floating bodies. When the haze lifts again, the dead bodies have mysteriously disappeared. Visualize this haze moving in and covering the pool and then moving up again. Picture the haze and you'll remember the 108th element, Hassium.

109.   Meitnerium

Meitnerium is element 109. *Meitnerium* sounds like "*meat in ear*". There's a good visual for you. Picture after the haze has lifted, there's a single body left lying by the side of the pool. Just a single body, and on closer inspection it looks healthy except for meat in its ear. Imagine a steak, a nice juicy beef steak hanging out of the ear of this body.

It has meat in the ear. "Meat in ear" will remind you of the 109th element, Meitnerium.

110. Darmstadtium

Next is Darmstadtium, the 110th element. That sounds like "*darned stadium*". On the body of the single body with meat in his ear, you notice a piece of string tied to his foot. You follow the colored piece of string and it leads to an entire darned stadium. Darning is what you do with holes in your socks. It's like knitting or crocheting and the entire stadium has been darned. It's like giant knitting has been done all over it and you can't see the stadium at all. Make it bright orange dots and pink stripes. "Darned stadium" will remind you of the 110th element, Darmstadtium.

111. Roentgenium

The 111th element is *Roentgenium*. We'll go with "*rent Jen*". You push through the darned curtain covering the entrance and go inside the stadium. It's a huge stadium but inside is just a lone person. It's Jen. Jennifer Aniston, and the entire stadium has been rented out to her. Rent Jen. "Rent Jen" will remind you of the 111th element, Roentgenium.

112. Copernicium

Finally, the last element, the 112th is Copernicium. *Copernicium*, named after Copernicus. "*Copper knees*" will be your image this time. You see Jennifer Aniston is the only person in the entire stadium and you walk closer. She's wearing a short skirt and because of the short skirt you can see both of her knees. They're made of copper. They're both reddish copper colored and when she walks, her knees knock against each other and make a clanging metallic sound. Picture those copper knees and you'll recall the 112th element, Copernicium.

Summary

And that's it!

One final revision for elements 101 to 112.

101. Mendelevium and you are reminded with an image of "mental leave". Einstein's firm body is taken by white men to a mental institution where he's put on mental leave. "Mental leave" equals *Mendelevium*.

102. In his sparse room, Einstein wakes up and looks around. He presses frantically at the call button for a nurse but there is no bell in the button and remains silent. "No bell" reminds you of *Nobelium*.

103. Instead of a nurse, Lawrence of Arabia walks into the room, reminding you of *Lawrencium*.

104. Lawrence is in the wrong room so he goes to the next room where his brother Ford is. Ford has a large Ford emblem tattooed in the center of his forehead. "Brother Ford" reminds you of *Rutherfordium*.

105. When Ford gets up out of bed, you see his legs have a zigzag bend because he has a double knee, one in the front and one at the back. Double knee or "dub knee" reminds you of *Dubnium*.

106. Ford grabs his C board, a special type of surfboard to accommodate his double knees, and heads to the pool. Focus on the C board and be reminded of *Seaborgium*.

107. When Ford gets to the pool, he talks to people but is such a bore they fall asleep listening to him and fall into the pool. Picture him being a bore and you will recall *Bohrium*.

108. A light fog or haze descends over the pool and covers all the floating bodies. When the haze lifts, all of the bodies have disappeared. The haze links in your memory to *Hassium*.

109. You see the only body left lying by the side of the pool looks healthy except for meat in its ear, a big steak. "Meat in ear" reminds you of *Meitnerium*.

110. A piece of colored string is tied to the body's foot and it leads to an entire darned stadium. "Darned stadium" reminds you of *Darmstadtium*.

111. Inside the stadium is a single person, Jennifer Aniston. It's only her because she has rented out the entire stadium for herself. "Rent Jen" reminds you of *Roentgenium*.

112. When you take a closer look at Jen, you notice she has copper knees which clang together. "Copper knees" remind you of *Copernicium*.

And that is the periodic table from 1 to 112.

14 UNUNTRIUM TO UNUNOCTIUM
(113-118)

On most periodic tables the elements go up to 118. From 113 to 118 the names are only temporary and have been designated as below. Until they are given permanent names we haven't included them in your memory story.

113. Ununtrium
114. Ununquadium*
115. Ununpentium
116. Ununhexium*
117. Ununseptium
118. Ununoctium

If you do wish to memorize the temporary names, the quickest way is to develop an understanding of their Latin derivations. Each name starts with "Unun-" and is followed by a suffix indicating their number.

"Tri" refers to 3.
"Quad" refers to 4.
"Pent" refers to 5.
"Hex" refers to 6.
"Sept" refers to 7.
"Oct" refers to 8.

*On May 30, 2012 the names Flerovium and Livermorium were adopted for elements 114 and 116, respectively.

However, because our memory story links the names of consecutive elements, we can't add Flerovium and Livermorium using the Link and Story method that we've used.

Here's a quick way you can memorize these extra elements.

All of the temporary names begin with 'Unun-' (Latin for 'one one') and are followed by a Latin suffix indicating their number.

"Tri" refers to 3.

"Quad" refers to 4.

"Pent" refers to 5.

"Hex" refers to 6.

"Sept" refers to 7.

"Oct" refers to 8.

If you start with 'Unun-' and add the Latin suffix, you'll have most of the name.

114.Flerovium

Flerovium is 114, which uses the Latin suffix 'quad'. Picture a square, which, like a quadrangle has four sides. Now make the square a bright fluoro color, like a *fluoro* pink. Fluoro will remind you of Flerovium.

116.Livermorium

Livermorium is 116, which uses the Latin suffix 'hex', like a hexagon. Your liver is the large organ on the right side of your belly beneath your lungs. Picture where your liver is, and imagine that you're getting a big tattoo of a hexagon right over your liver. Visualize that hexagon over your liver, and you'll remember that Livermorium is element number 116.

And that's all there is to it, now you've memorized the final five elements of the periodic table!

15 REMEMBERING ATOMIC NUMBERS

Congratulations, you've now mastered the periodic table from the first element through to the 112th element.

The mnemonic technique we've used in this book has allowed you to memorize the names of the chemical elements easily and without learning a memory system. And that was our goal.

However, the way the Link and Story Method works means you can only recall the elements in order. If someone asks you what the 24th element is, you'll need to work through the elements from 1 to 24 to confirm it is Chromium.

There are mnemonic systems which would allow you to instantly know the atomic number for any element, but those systems are more complex and require more effort to learn. Instead, we'll go for a compromise.

In this chapter you'll learn an expert memory system called the Rhyming Peg System which is simple to use. Using this method, you'll memorize which elements have the atomic numbers 10, 20, 30 and so on. Then, when you're asked what the 24th element is, you'll be able to instantly recall the 20th element, go through your visual memory story from there and quickly arrive at Chromium.

Alternatively, if someone asks what atomic number Chromium has, you'll be able to begin your visual memory story at Chromium (the doctor made of shiny chrome) and continue the story until you recognize Zinc (the kitchen sink) is 30. From there you can quickly calculate Chromium is 24.

Let's learn the Rhyming Peg System, and then we'll adapt and apply it to the periodic table, and practice using it.

Rhyming Peg System

Like all expert memory techniques, the Rhyming Peg System is based on mental imagery. Every number from one to ten has an image associated with it. The name of the image rhymes with the name of the associated number, making it simple to learn and remember.

One = Gun
Two = Shoe
Three = Tree
Four = Door
Five = Hive
Six = Sticks
Seven = Heaven
Eight = Gate
Nine = Wine
Ten = Hen
Eleven = Leavened (bread)

Rhyming pegs typically only go to ten, but you'll understand shortly why we're including an image for eleven too.

Go through the list of numbers and images, and clearly visualize a picture of each image. One is gun, so imagine a large, shiny hand gun like you see in the movies. Two is shoe. Picture an enormous basketball shoe the size of a car. Create your own mental image for each rhyming peg word.

If you work through the entire list a couple of times, before long you'll be able to 'count' using the rhyming peg words. Gun, shoe, tree, door, hive, sticks, heaven, gate, wine, hen, leavened.

You've now learned the basics of the Rhyming Peg System. Next, we'll adapt it to the periodic table.

We're going to add each of the rhyming peg images to a related chemical element. There are 112 elements and we have eleven rhyming images, so we'll only link a rhyming image to every 10th element.

Gun (one) will be linked to the 10th element. Shoe (two) will be associated with the 20th element. You'll link tree (three) to the 30th element, and so on. Just add a zero to the rhyming peg number you'll have the related element number.

In addition, when you add each of the rhyming images to your existing memory story, you'll picture them as being on fire. That will make the images stand out in your mind and you'll realize they're 'marker' images, and the related element must be number 10, 20, 30 etcetera.

Let's start adding the rhyming images to your memory story, and you'll quickly understand how it works.

10. Neon

We'll begin with the 10th element, neon. You can remember from your memory story the "night row general" has a special type of fluoride on his teeth. When he smiles his teeth glow like neon lights, prompting you to recall the 10th element, Neon.

Into that mental picture, you're going to add an image of a gun. One rhymes with gun. Picture the "night row general" and imagine he's wearing a giant hat on his head, and it's in the shape of a gun. Weird, right? When he smiles and his teeth glow neon, the gun hat suddenly bursts into flames. It's like a light switch. Imagine what he looks like when he smiles and then stops smiling. His teeth glow neon and the large gun on his head bursts into flames, and when he stops smiling they quickly turn off.

Clearly picture the giant gun being worn as a hat, and imagine it bursting into flames at the same time as the general's teeth glow neon.

When you're asked which the 10th element is, you'll think of the rhyming peg for one – gun. An image of the large flaming gun on the general's head will appear in your memory, and you'll see the neon teeth glowing in unison with the flaming gun. And you'll immediately know the 10th element is Neon.

20. Calcium

In our rhyming peg system, two is shoe, and you'll link shoe with the 20th element – Calcium. You can remember calcium because of the miniature coliseum. Inside the coliseum are the crowds of screaming spectators who want the blood of the gladiators.

To that image you'll add a shoe. Floating above large sporting events you often see blimps, commonly used for advertising. Above the mini Colosseum imagine a huge blimp in the shape of a shoe. And it's on fire. Instead of seeing the Goodyear blimp floating across the sky above the Colosseum, it's an enormous flaming shoe. Like a fireball slowly drifting across the sky.

Now, when you think of calcium, you'll picture the Colosseum and above it you'll imagine the giant shoe floating along, trailing bright orange flames and smoke. Because shoe rhymes with two, you'll know Calcium is the 20th element.

30. Zinc

We're up to the 30th, so we need the rhyming peg for three – tree. The 30th element is zinc and you remember it by picturing a zinc kitchen sink being kicked by a Viking underwater.

Clearly picture the Viking kicking the zinc sink with his toe again, and now imagine when he kicks the sink, an underwater tree next to the sink suddenly lights up with roaring flames. The tree is enormous and it's growing on the bed of the ocean or river. When the Viking kicks the sink, it's like a light switch being turned on, and the huge tree erupts into flames for a brief second.

Link your mental picture of the flaming tree to the kitchen sink, and you'll associate it with the element zinc. Because tree rhymes with three, you'll know Zinc is the 30th element.

40. Zirconium

In the rhyming peg system, four is door. Add a zero to four and you know you're linking the 40th element to door. The 40th element is Zirconium, which you remember because of thousands of fir cones falling to the ground and blanketing the area.

To that scene of thousands of fir cones covering the ground you'll add an image of a flaming door. Imagine that expanse of fir cones, and in the middle of it all is standing a door. It's completely unsupported, there's no doorframe or wall, it's just a door standing up by itself. And it's on fire. There's thick black smoke pouring off the door as it burns brightly.

When you want to recall what the 40th element is, you'll know the rhyming peg for four is door. An image of a brightly burning door with thick black smoke will appear in your memory, and you'll see it's standing in the middle of thousands of fir cones. The fir cones will be associated in your mind with Zirconium.

50. Tin

The 50th element is Tin, but how will you remember it's number 50? You'll link your memory story for tin to an image of hive, the rhyming peg for five.

In your visual story, the image for tin was the frightened golf caddie throwing a tin can at the angry Indian lady. Picture again that tin can flying through the air, as if it was in slow motion. Imagine it flies straight over a hive of bees. The hive itself is on fire and thousands upon thousands of swarming bees are escaping the burning hive. The tin can flies straight through the middle of the swarm.

When you clearly visualize that tin can flying over the burning hive of swarming bees, you'll link hive to five. And once you've added a zero to five it becomes 50, and you'll instantly recall the 50th element is Tin.

60. Neodymium

The rhyming peg for six is sticks, which will prompt you to recall the 60th element, Neodymium.

To remember Neodymium you imagined Neo, Keanu Reeves' character in *The Matrix*, walking into a tree. Animated cereal cry "Neo dim, Neo dim" because it was not a demonstration of Neo's smarts. Now you have to add some sticks to that visual scene.

Let's make the sticks drumsticks, and of course they're on fire. While half of the cereal are chanting "Neo dim, Neo dim", the other half start to beat large bass drums with flaming drumsticks. There's a deafening rhythm of drum beats accompanying the chant of "Neo dim". Picture this primal image and the sound of the chants, and smell the smoke from the burning drumsticks as they beat a heavy rhythm.

When you need to recall the 60th element, you'll think of six. Six rhymes with sticks, and you'll visualize the burning drumsticks accompanying the chant of "Neo dim", which you'll associate instantly with Neodymium.

70. Ytterbium

Moving on to 70. The rhyming peg for seven is heaven, and the 70th element is Ytterbium.

The substitute phrase for Ytterbium is "hit herbs". You use a tool to hit herbs while you're standing in the middle of them, flattening as many herbs as you can. To this image you'll add a picture of heaven, or the sky and the clouds.

Every time you hit the herbs with a thud, there's a bright flash in the sky as the clouds all burst into flame. Visualize yourself swinging the tool down on the herbs, and as it makes contact there's a blinding light in the heavens as the clouds flash with flame. It's incredibly spectacular and dramatic.

To recall the 70th element, your mind will leap from seven to heaven, and then to an image of the heavens flashing with fire as you hit herbs. "Hit herbs" leads you directly to Ytterbium, the 70th element.

80. Mercury

Eight rhymes with gate, and will link to number 80, which is Mercury. You need to add an image of a flaming gate to your memory story for Mercury. To recall mercury you imagined a flood of thick, silver liquid mercury. Keep picturing that liquid mercury flood and imagine it pushing through two enormous gates. Make them really huge, like they'd normally fit on the front of a castle. As soon as the liquid mercury touches the gigantic timber gates, they burst into flame and blaze away. Once again, make this image quite dramatic and spectacular in your mind.

Now, when you need to recall the 80th element you'll think of the number eight, rhyme that with 'gate', and picture the enormous gates being set on fire by the flood of liquid mercury. And you've successfully remembered the 80th is Mercury.

90. Thorium

How will you memorize Thorium is the 90th element on the periodic table? First you'll recall the rhyming peg for nine is wine, and then you'll create a visual association of wine and Thor.

Your memory story for Thorium had Thor, the God of thunder and lightning, smashing the city of Paris with his hammer. He wanted to get rid of all the actors, which he succeeded in doing. Now imagine while Thor is busy smashing the city with his hammer, he's also casually drinking wine straight from a bottle. After each mouthful he exhales and breathes fire in long shooting flames. He's like a fire breathing dragon, as if he wasn't scary enough with his destructive hammer.

When you recall the fiery wine, you'll know it's linked to the 90th element and your mental picture of Thor will remind you of Thorium.

100.   Fermium

In your visual memory story you associated the element fermium with an image of Einstein's entire giant body becoming firm and turning into a statue. Fermium is the 100th element and you'll now add to that image by including a hen. Hen rhymes with ten.

As we know, statues are a favorite resting place for birds. Imagine on the shoulder of the Einstein statue is sitting a hen. It's a large hen with pure white feathers and a scarlet red crest. The hen is sitting peacefully on the statue's shoulder and appears unaware its tail feathers are on fire. You can see the flames rising up above the head of the hen, and beginning to singe the hair of the Einstein statue.

When you want to remember the 100th element you'll jump from ten to an image of a flaming hen. Your clear mental picture of the flaming hen sitting on the shoulder of the completely firm Einstein will come to mind, and you'll know the 100th element is Fermium.

110.   Darmstadtium

Finally, Darmstadtium is the 110th element of the periodic table. You'll recall from the memory story the visual association is an entire stadium covered in darn, like a woolen jumper. That needs to be linked to the rhyming peg for eleven, which is leavened bread.

Unleavened bread is flat, thin bread because it doesn't have yeast in it to make it rise. For the sake of our story, let's say leavened bread is the complete opposite. It has yeast and rises and becomes a high and fluffy loaf.

All over the darned stadium are hanging thousands of loaves of leavened bread. They're hanging by pieces of wool that's come loose from the darning. And all of the leavened bread has been badly burned, and many of them are slowly breaking into flames. There's the familiar smell of burned toast in the air, and you're concerned the flames will cause the darned stadium to catch light. Don't worry, wool is quite fire resistant.

When you're quizzed on what the 110th element is, you'll know recall eleven rhymes with leavened bread, and you'll picture all of those high, fluffy loaves burning as they hang from the darned stadium. The darned stadium will remind you immediately of Darmstadtium.

You're now able to remember the 10th, 20th, 30th, 40th, 50th, 60th, 70th, 80th, 90th, 100th and 110th elements.

How do you recall what the 53rd element is? You start by remembering the nearby 50th element. Five rhymes with hive. You visualize the burning hive and a tin can flying through the swarm of escaping bees. So the 50th element is Tin. From there you continue through your full memory story. The tin can is sold in exchange for ant money – antimony is 51. The teller at the bank won't accept the ant money –Tellurium is 52. The Indian cries "I owe Dean" – the 53rd element is Iodine.

And how will you know what atomic number Copper is? You'll be reminded of your memory story where copper was represented by copper pipes. They were being used by Vikings diving underwater. One of the underwater Vikings kicks his toe on a zinc kitchen sink, and a large tree growing nearby bursts into flames. The flames indicate zinc is a 'marker' element. It's the tree that's on fire and tree rhymes with three, indicating the 30th element. If Zinc is 30, Copper must have an atomic number of 29.

Summary

It's time for a quick summary of the rhyming peg images and their associated elements.

First, the rhyming peg images; one is gun, two is shoe, three is tree, four is door, five is hive, six is sticks, seven is heaven, eight is gate, nine is wine, ten is hen and eleven is leavened bread.

When you add a zero to each of the numbers, gun becomes the marker image for the 10th element. Shoe is the image for the 20th element, and so on. Each of the images is on fire in some way, making them stand out in your mind.

10. The <u>gun</u> is on the head of the 'night row general', and it bursts into flames in unison with his teeth glowing neon. *Neon* is the 10th element.

20. The <u>shoe</u> is a giant burning blimp, floating over the miniature Colosseum. *Calcium* is the 20th element.

30. The <u>tree</u> flashes with flames when the kitchen sink is kicked by the underwater Viking. *Zinc* is the 30th element.

40. The <u>door</u> is burning and smoking while standing by itself in the middle of thousands of fir cones. *Zirconium* is the 40th element.

50. The burning <u>hive</u> is beneath the swarm of bees, through which the tin can flies. *Tin* is the 50th element.

60. The flaming drum<u>sticks</u> are being wielded by animated cereal as an accompaniment to the chants of "Neo dim". *Neodymium* is the 60th element.

70. The <u>heavens</u> flash with flame each time you hit herbs. *Ytterbium* is the 70th element.

80. The <u>gates</u> catch fire as the flood of liquid mercury touches them. *Mercury* is the 80th element.

90. The <u>wine</u> is being drunk from the bottle by the fire breathing, hammer wielding Thor. *Thorium* is the 90th element.

100. The <u>hen</u> is blissfully unaware its tail is on fire as it sits on the shoulder of the firm, giant Einstein. *Fermium* is the 100th element.

110. And thousands of loaves of <u>leavened</u> bread hangs from the darned stadium, some burned already, others still on fire. *Darmstadtium* is the 110th element.

A Final Word

You have now memorized the entire periodic table in the quickest and easiest way possible. Did you ever think your memory was so incredible?

To ensure your visual associations are transferred permanently to your long term memory, try to review your memory stories five times over the coming days, weeks and months. It will be the easiest revision you've ever done, and the most enjoyable.

If any of the element names remain elusive, revisit the substitution images and story and focus on creating a truly vivid picture of them in your mind. And then trust your memory to do the rest.

We hope you've enjoyed this innovative new way to learn and memorize as much as we've enjoyed creating it.

Always remember, you have an amazing memory.

ABOUT THE AUTHORS

Kyle Buchanan and Dean Roller specialize in memory training and innovative educational products, and are the founders of Memory Worldwide Pty Ltd.

They co-authored *"Memorize the Periodic Table: The Fast and Easy Way to Memorize Chemical Elements"*, which quickly became the #1 best-selling book in Science Education on Amazon.com. Their latest venture is MemorizePeriodicTable.com, which helps students and adults memorize the entire periodic table with the use of animated videos, an exciting new innovation in education and memorization.

With six university degrees between them, Kyle and Dean understand the process of research and learning. Their latest research focus is in the fields of educational and cognitive psychology as they study and write about memory mnemonic techniques and their practical application.

Kyle and Dean's goal is to help you learn and memorize new information faster and more easily than you ever imagined possible.

www.ingramcontent.com/pod-product-compliance
Lightning Source LLC
Chambersburg PA
CBHW061836220326
41599CB00027B/5295